My Body and Me

Author - Larmie Mowry

Illustrations - Nicole Arthur

Tucson, Arizona

MYBODYANDMECOLOR

This Book Belongs To:

_____

MY BODY AND ME - SELF PORTRAIT!

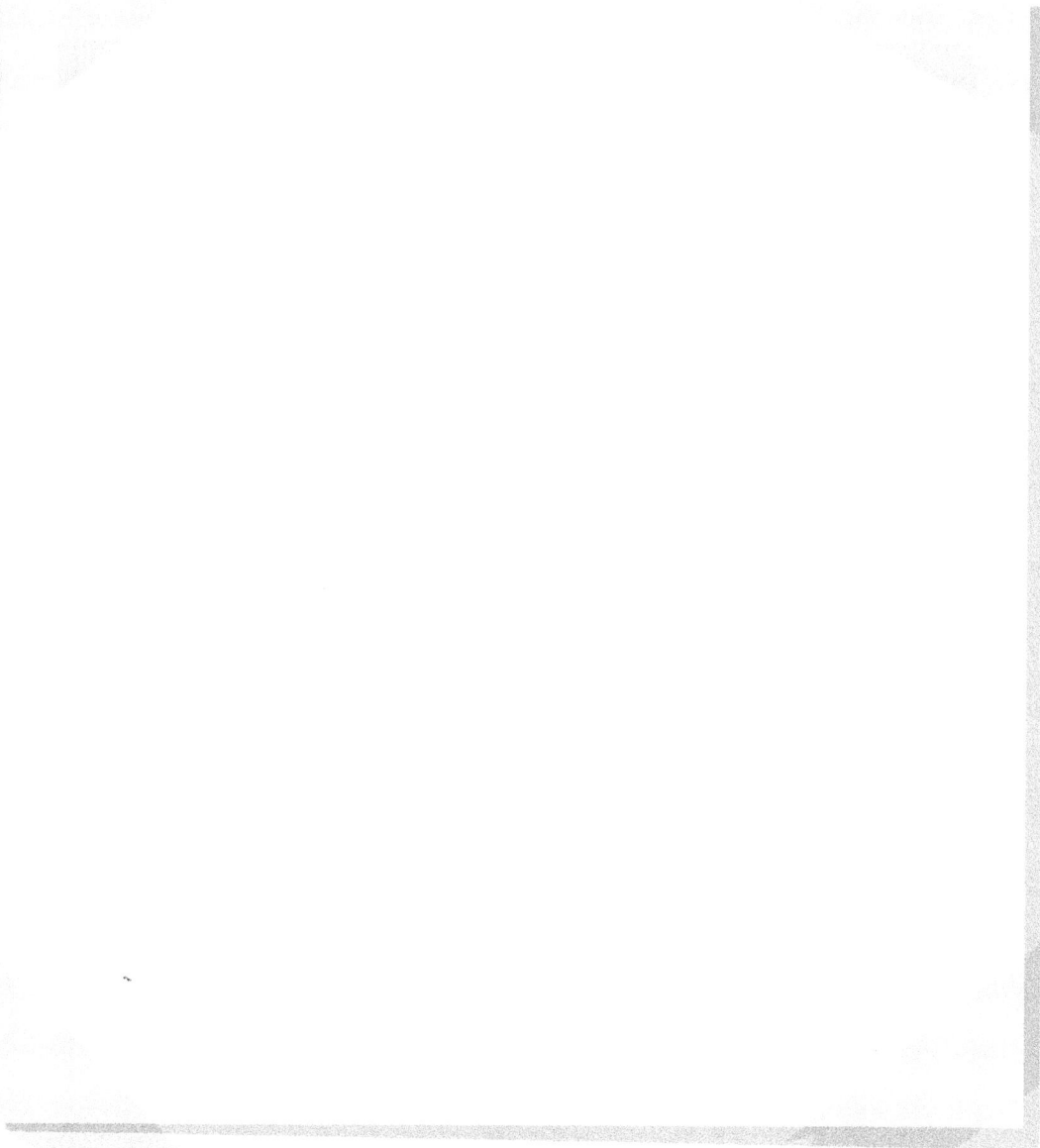

Hi! My name is Nelson and I am your Nervous System! Even though I am your nervous system I am actually not nervous at all! I am called that because a big part of me are something called nerves! The other parts of me are your brain and your spinal cord. All of my parts work together to help you think, see, hear, smell, taste and feel. I have a really big job to do which is why I can be found all throughout your body.
Isn't that cool?

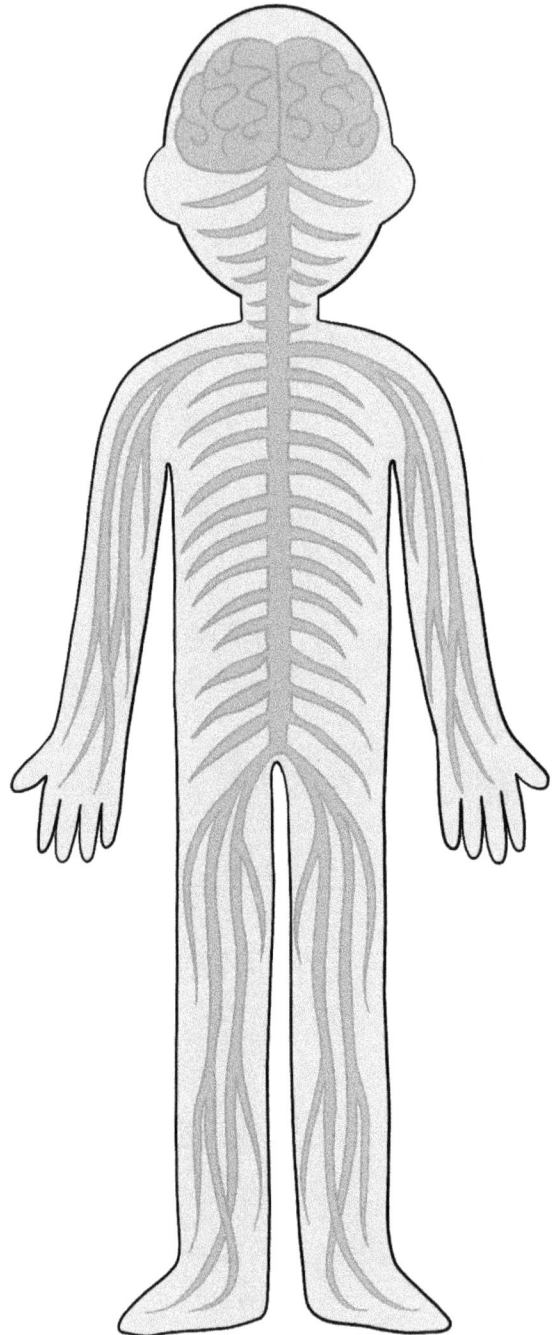

N

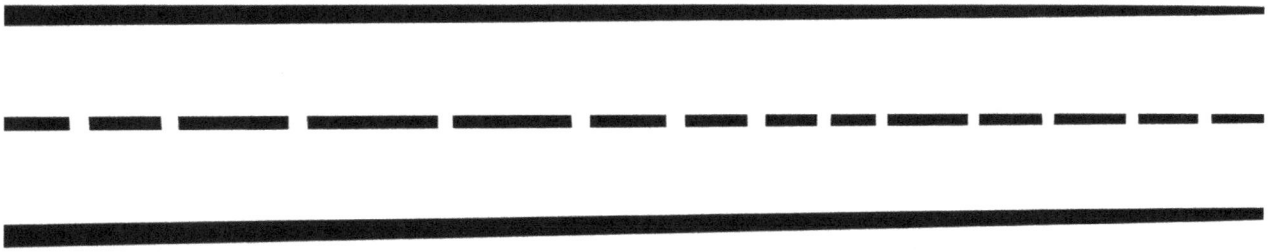

For Nervous
System

Hi! I am your brain and I am in charge of your whole body! I have a really big job and it takes a lot work. I am always growing and always learning, which means you are always growing and learning! The brain has a lot of different parts that are in charge of a lot of different things like talking, hearing, moving, thinking, memory and more!

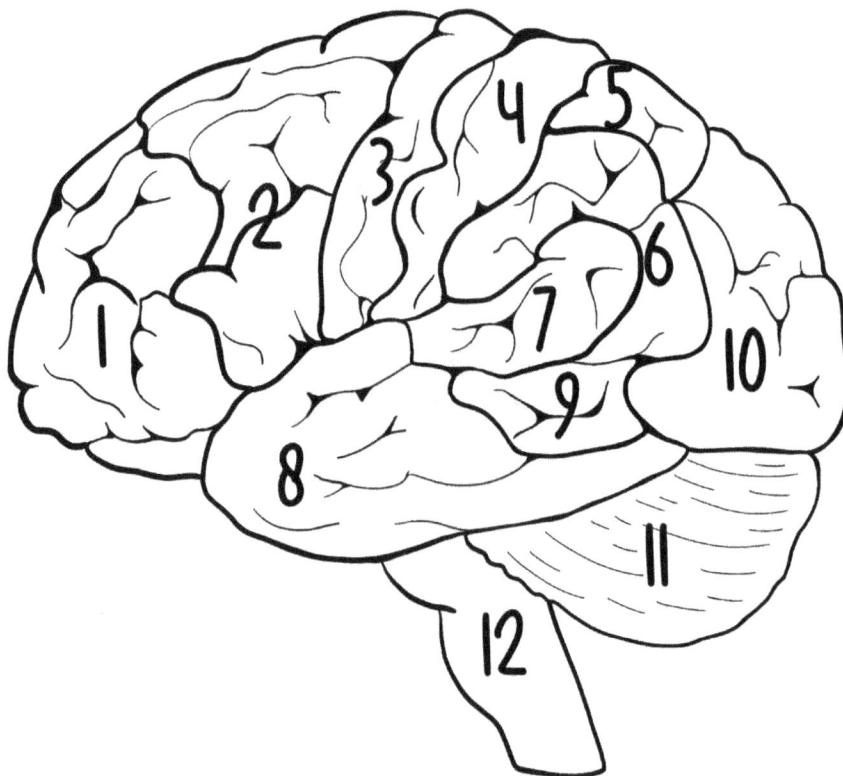

1- Purple
2- Yellow
3- Dark Green
4- Red
5- Pink
6- Dark Blue
7- Light Orange
8- Light Blue
9- Grey
10- Dark Orange
11- Light Green
12- Black

Hi I am your spinal cord! I am protected by the bones in your spine. I talk to all of your nerves and neurons and tell the brain what they are saying. We all work together to make sure we do our best to keep you safe!

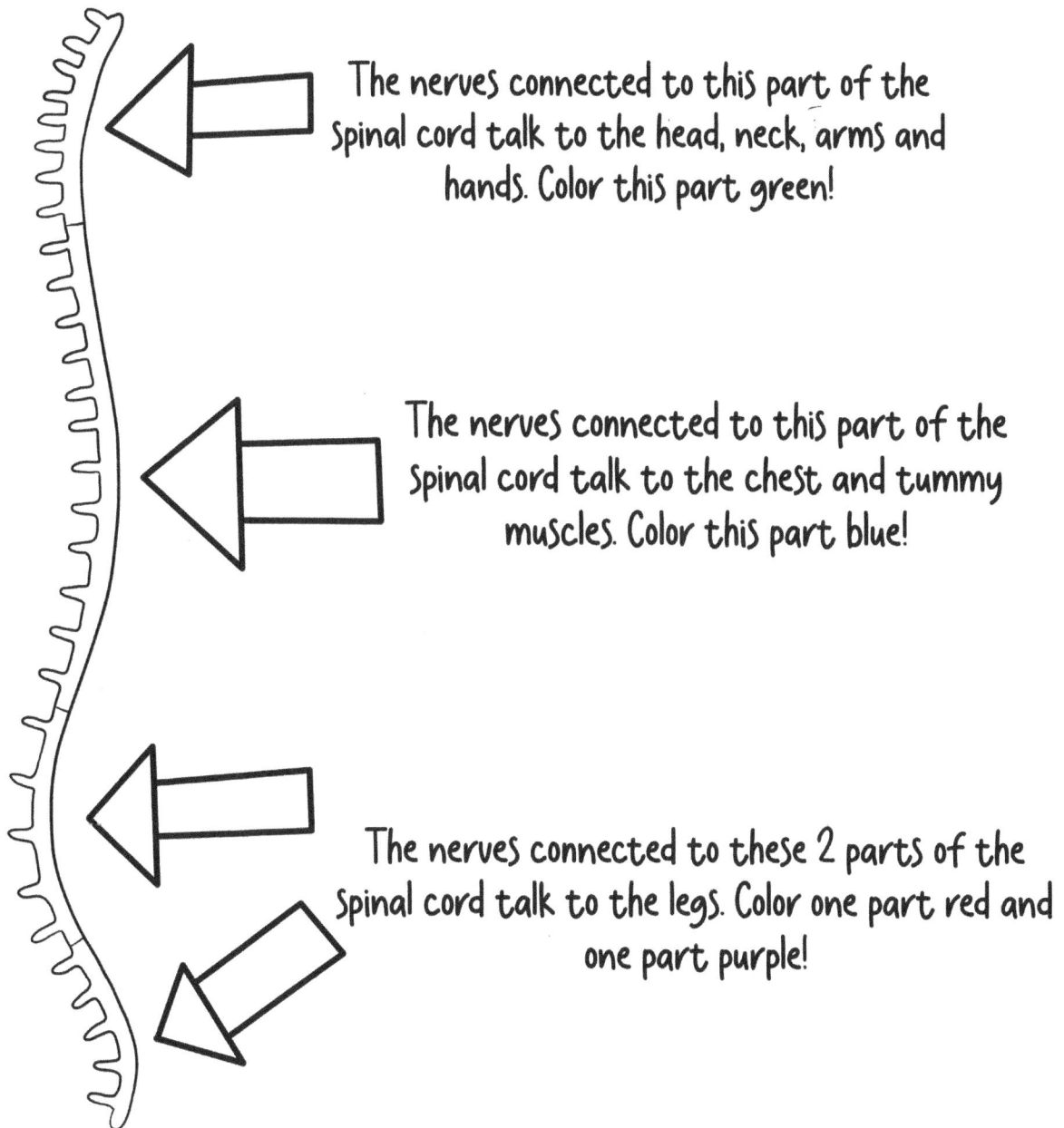

The nerves connected to this part of the spinal cord talk to the head, neck, arms and hands. Color this part green!

The nerves connected to this part of the spinal cord talk to the chest and tummy muscles. Color this part blue!

The nerves connected to these 2 parts of the spinal cord talk to the legs. Color one part red and one part purple!

Hi! I am a neuron and I am the smallest piece of the nervous system. I talk to all the different parts of your body and then talk to your spinal cord which talks to your brain.

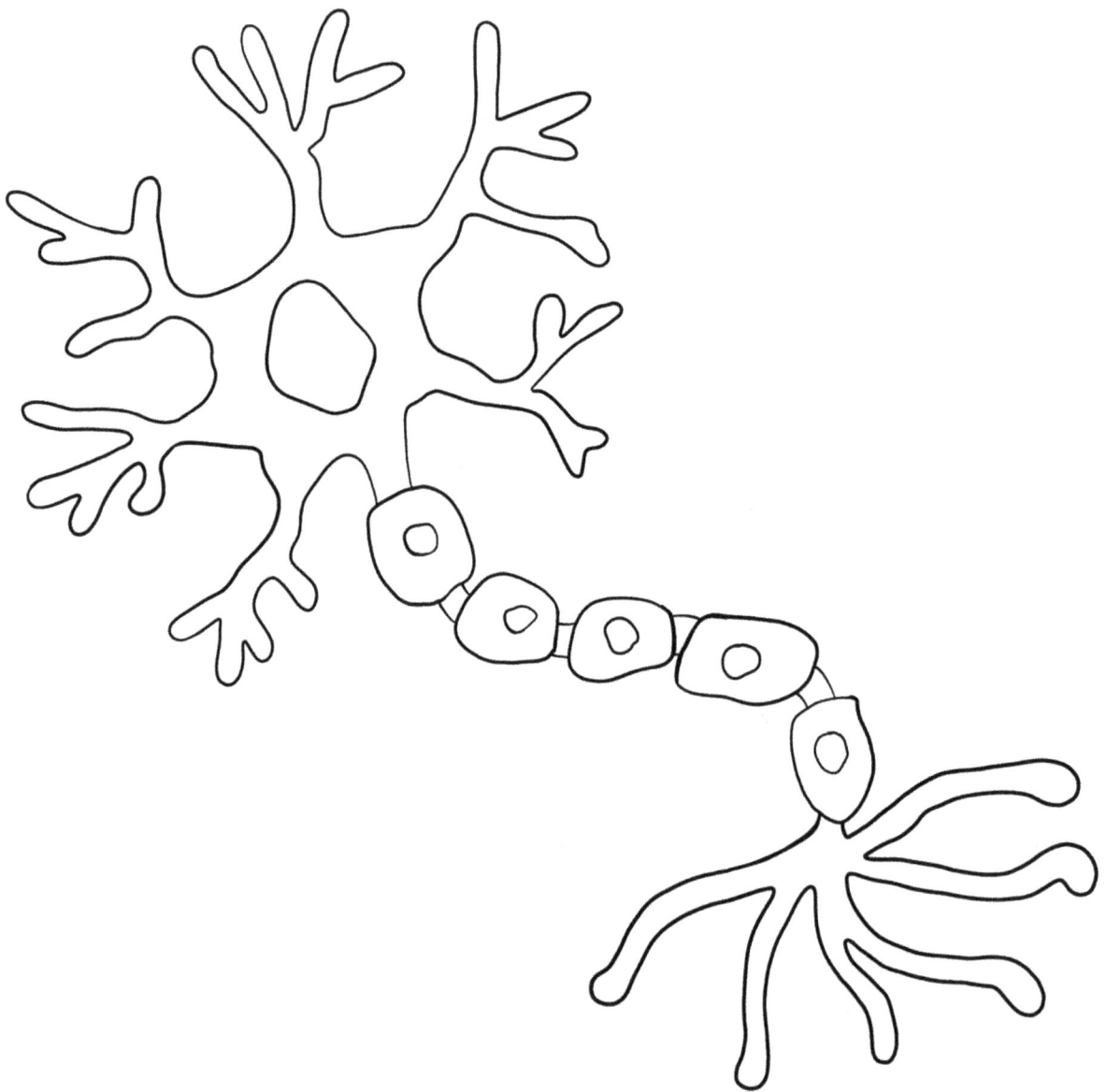

This is an eye and it's job is to let you see. Your eye talks to your brain and your brain tells your eyes what they are seeing. What color are your eyes?

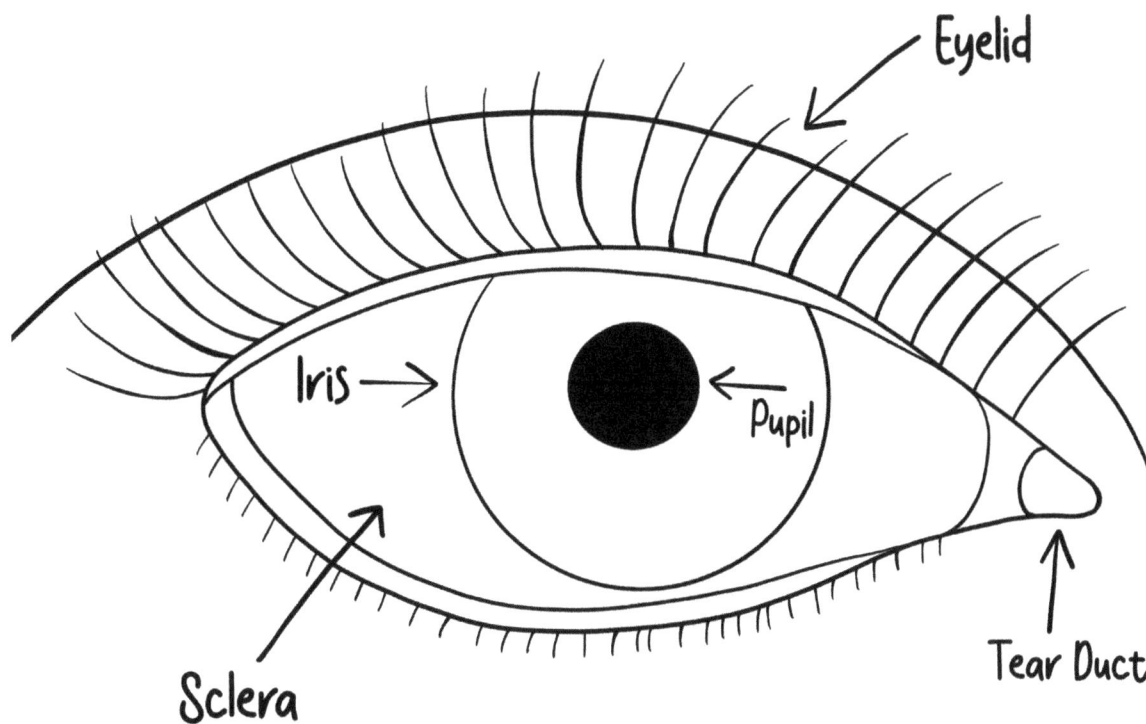

Eyelid

Iris →

← Pupil

Sclera

Tear Duct

Your eyelid is the outside part of your eye and helps your eye open and close.
Your tear duct is in charge of making tears when you cry.
Your sclera is the white part of your eye and it protects your eye from injury.
Your iris is the colored part of your eye and helps you see clearly.
Your pupil is the black part of your eye and it lets light into your eye.

This is your ear and it's job is to hear all of the sounds around you! Your ear hears the sounds then talks to the brain and the brain tells you what you hear.

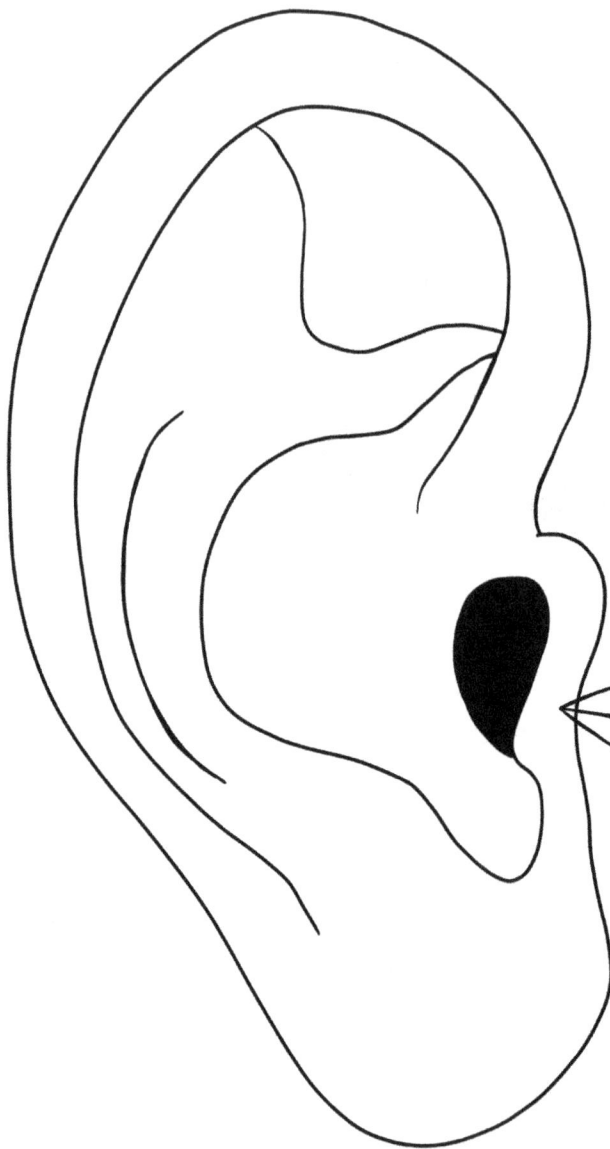

This is your ear canal and that is where sounds get in and travels to your brain

Color then match the pictures to show how well you can see!

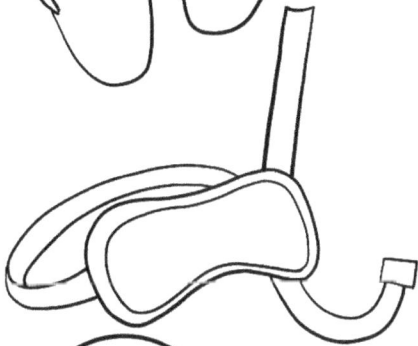

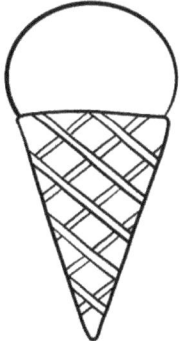

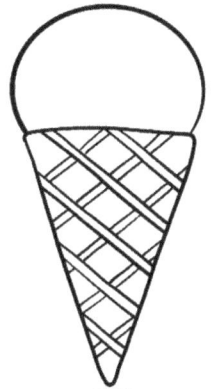

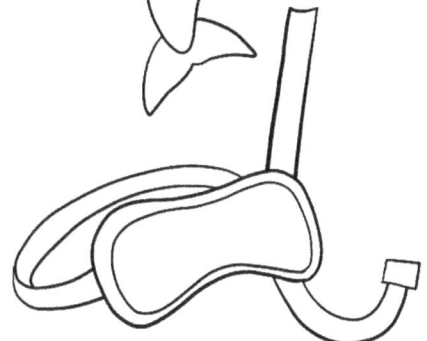

What kind of sounds do these pictures make?

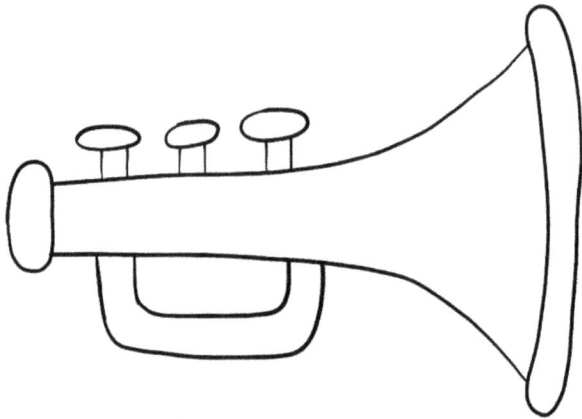

This is your mouth and it's job is to help you talk, taste, breathe and eat. Your tastebuds talk to your brain and then your brain tells you what you are tasting and whether you like it or not.

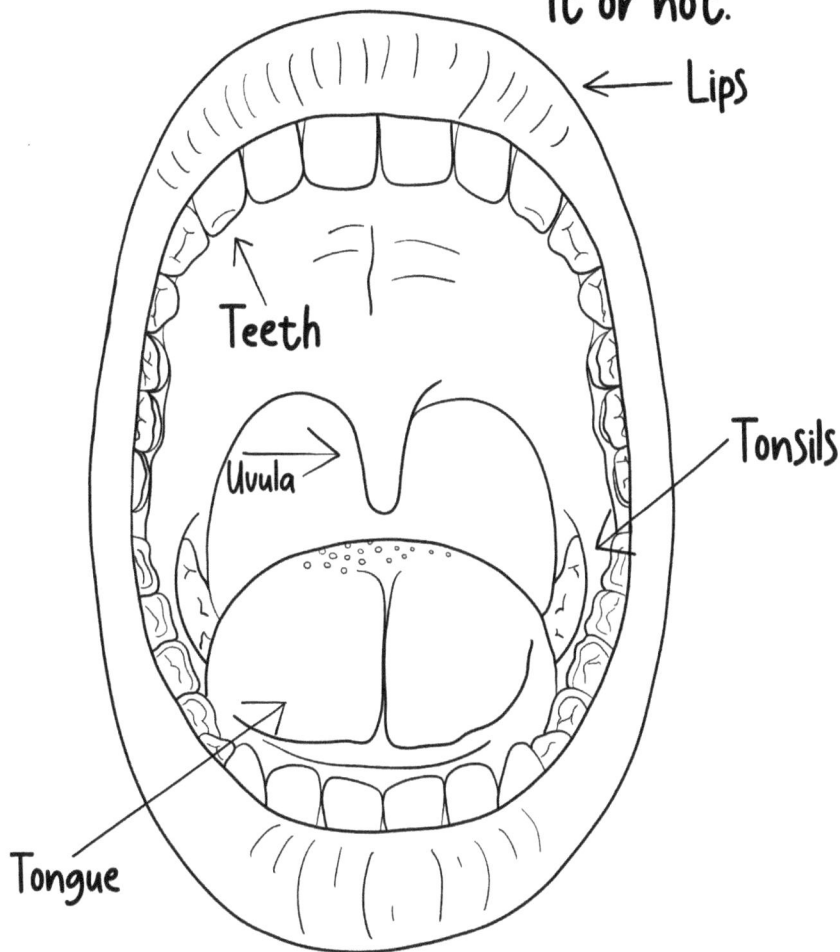

Lips

Teeth

Uvula

Tonsils

Tongue

Your lips are muscles that help you talk and keep your mouth closed.
Your teeth are bones in your mouth that crush your food so you can swallow it and not choke.
Your tongue helps you talk and eat. It also has little bumps called tastebuds that tell you what stuff tastes like.
Your tonsils help keep you helping by stopping germs from getting into your body.
Your uvula dangles in the back of your throat and it helps your food and drink go down your throat instead of into your nose when you swallow

Your tastebuds tell you what you like and don't like. Which of these foods do you like?

Draw your favorite food and your least favorite food

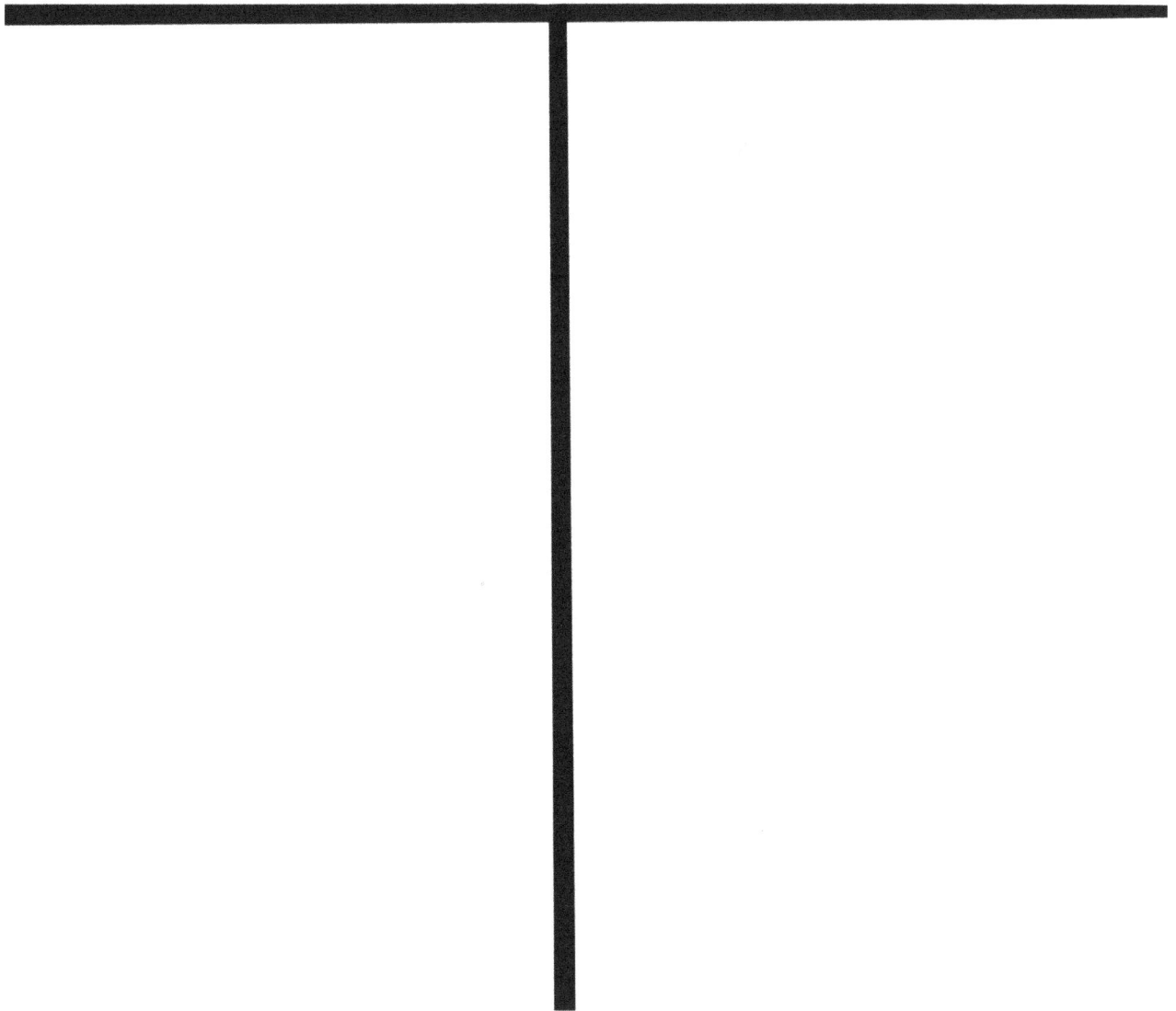

This is your nose and it's job is to help you breathe and smell. Your nose talks to your brain and then your brain tells you what you are smelling.

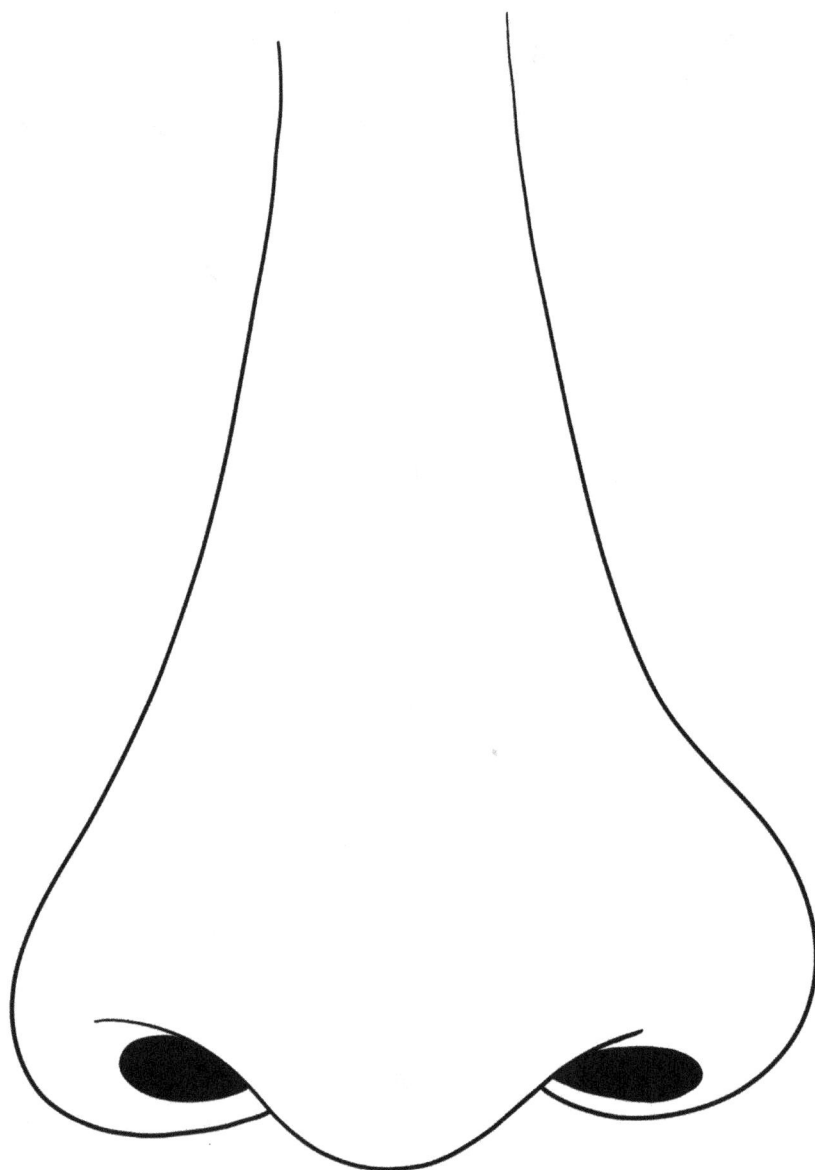

Color these items that you can smell

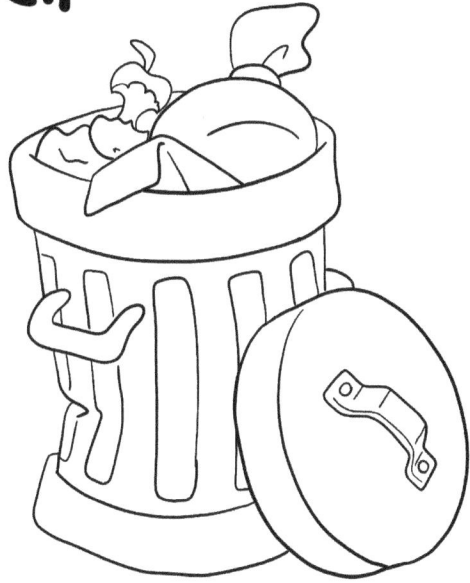

This is your hand and it has so many jobs! The hand helps you feel different textures, hold objects, push, pull draw and more!

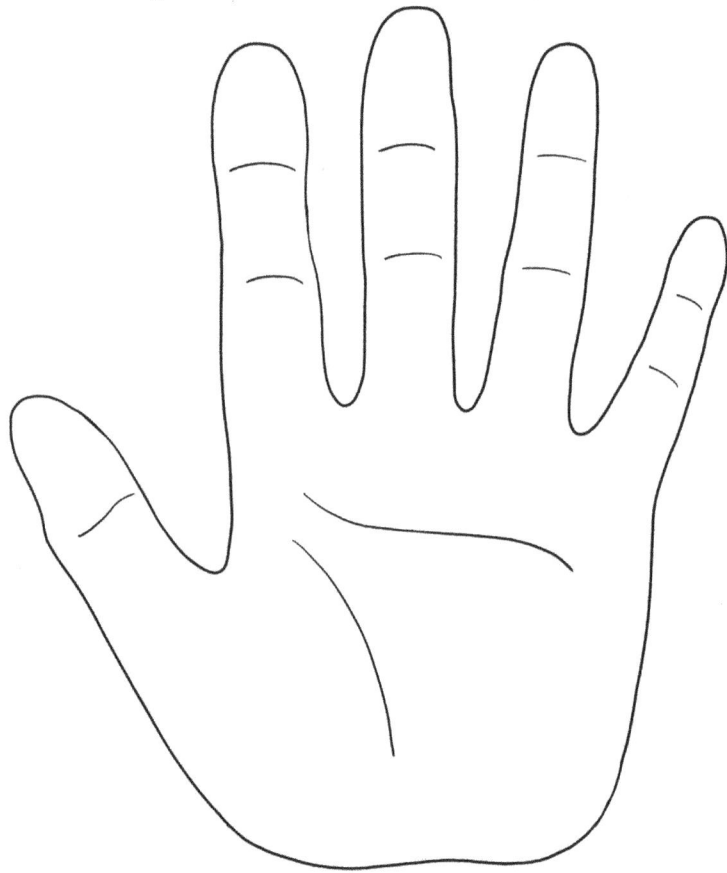

Your hand has fingers, knuckles, palms, and fingernails.

Color these things that all feel different

Every finger has a fingerprint and every person has different fingerprints.

Start

End

How many senses do you have?

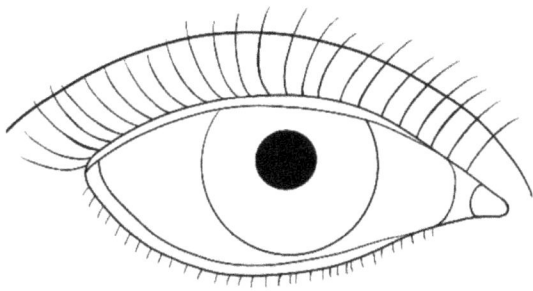

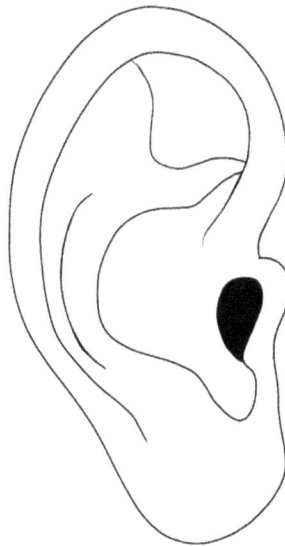

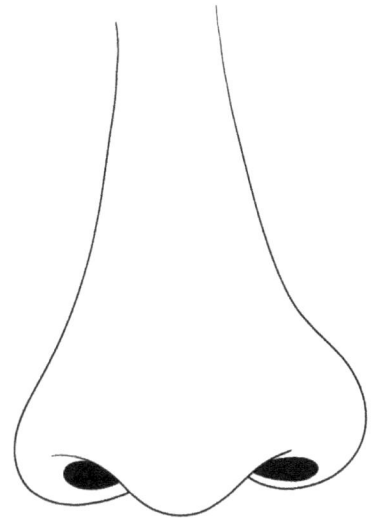

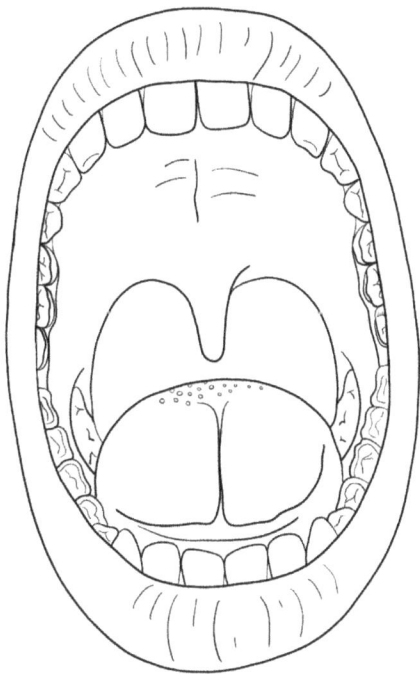

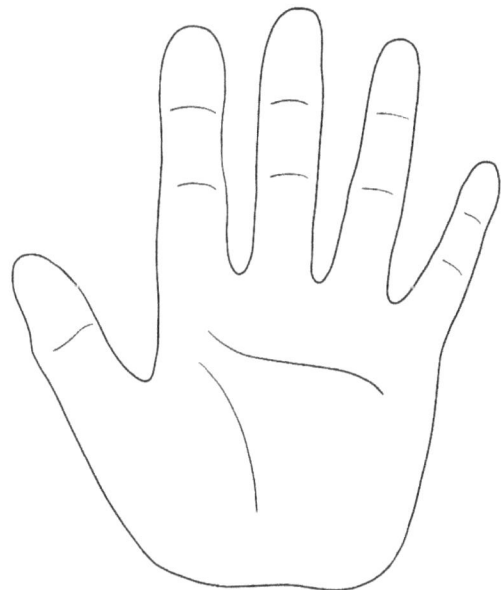

Your nervous system is also in charge of pain. Your nerves are what tells you if something is hot, cold, feels good or is painful. When something hurts it's not fun, but there are lots of ways to help!

Color the picture to find the hidden word!

1- Green
2- Purple
3- Orange
4- Pink
5- Red
6- Yellow
7- Blue
8- Brown
9- Grey
10- Black

Sometimes pain doesn't need to be fixed with medicine. Here are some ways to help when you have pain.

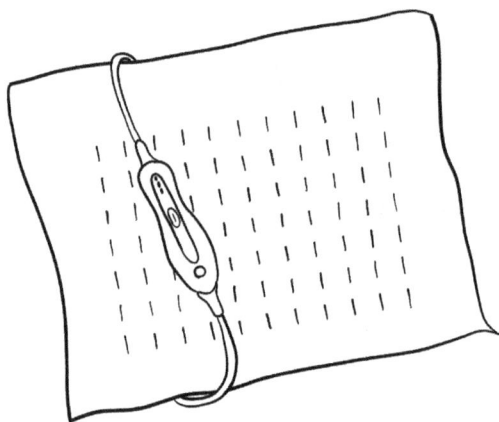

Draw your
favorite show in
the TV

www.ingramcontent.com/pod-product-compliance
Lightning Source LLC
Chambersburg PA
CBHW051350290326

41933CB00042B/3354